des&desi®

# MATTER Is All Around You

**WRITTEN BY TERRESSA BOYKIN**
**illustrated by Allison Michas**

ISBN 13: 978-1-945751-07-3

*Library of Congress Cataloguing-in-Publication Data*
2016962986

Boykin, Terressa
des&desi
*Matter Is All Around You*

**DEDICATION:**

This book is dedicated to Tracey Witten Rose because good friends truly MATTER.

Look inside your bedroom.
What things do you see?

All of the things around you
are the things **MATTER** can be –

The bed, the chairs, and pillow too,
your teddy bear and more,

The books you color in each day,
with crayons left on the floor.

COLORING
BOOK

MATTER comes in different shapes –
circles, squares, and lines.

It comes in many colors,
yellow, red, and blue designs.

MATTER is the many fruits we eat
like bananas, grapes, and citrus treats

And vegetables like kale and yams
or carrots, corn, and beets.

MATTER is rolling wooded hills
or the grass and rocks below,

And all of the animals in the forest
fury critters, a buck, and his doe.

On farms, it's animals of
different sizes – horses, pigs, and cows.

It's animals in the rainforest –
snakes, monkeys, and even owls.

It's falling leaves in an autumn breeze,
with branches hanging low,

And frosty icicles in the winter months
and snowflakes in the snow.

In the Spring and Summer months,
with the scent of flowers everywhere,

As the warm breeze passes by
with dandelions in the air.

But these things are the MATTER
that we see only with our eyes.

There is more to MATTER on the
earth and high above the skies.

When MATTER is gigantic
and really far away,

A telescope is best to see
the MATTER in this way.

With it we see the **MATTER** of
the moon, planets, and stars –

Neptune, Venus, Mercury, Saturn,
Uranus, Jupiter, and Mars.

And when MATTER is really small
like an insect in the grass,

We can use a tool to help us see;
it's called a magnifying glass.

But, what if this MATTER is
tinier than a spec of dust?

If our eyes cannot see it
then we need a tool we can trust.

To see MATTER that's so small
we may not know it's there.

A microscope is what we use
to see what may look bare.

All of these things are MATTER,
whether they're big or small

And, if you spent one thousand years
you could never name them all.

Yes, MATTER is all around you
in everything you see,

Inside, outside, and up above,
yes, MATTER is you and me!

MATTER is Everything!!!!

CHOOL

THE END.

# ACKNOWLEDGEMENTS


## ABOUT THE AUTHOR

Terressa Boykin holds a Masters of Science in Organic and Kinetic Chemistry, has worked in the field of Polymer Chemistry and Chemical Engineering for over 28 years. In addition to volunteering in her community as a local and regional science fair judge, she is a member of the American Chemical Society's Coaching Program, the Education and STEM Committee for her local Chamber of Commerce and is the founder and CEO of Des&Desi Chemistry Classes for young children and the author of the children's book series Des&Desi *Chemistry In My Dreams, Chemistry In My Dreams Coloring Book 1: Natural Processes, The Atom* and *Matter Is All Around You.*

www.desndesi.com

## ABOUT THE ILLUSTRATOR:

Allison Michas is a high school student and a self-taught artist. She likes to incorporate her love of science into her artwork. Digital painting is one of her favorite mediums, although she loves traditional drawing, painting, and sculpture as well.

## MY INSPIRATION:

Desmen A. Boykin is a middle school student who has a passion for science and a talent for creative drawing. He hopes to be an engineer one day.

Desiree M. Mosley is an LPN and currently an RN nursing student who has a gift and a passion for helping others.

# Other books by the author

**DES&DESI**

## CHEMISTRY IN MY DREAMS

**BOOK 1: NATURAL PROCESSES**

**DES&DESI**

## CHEMISTRY IN MY DREAMS

**BOOK 2: MAN-MADE PROCESSES**

DES&DESI

## THE ATOM

### WHAT AM I REALLY?

DES&DESI

## CHEMISTRY IN MY DREAMS

### COLORING BOOK 1: NATURAL PROCESSES

Made in the USA
Monee, IL
05 November 2026

67385076R00026